Vision

Write. Plan. Pray

A Strategic Trilogy Designed to Bring Your Vision to Life!

Dr. Kimberly Ellison

Printed in the United States of America

First Printing, 2016

ISBN-978-0-692-81451-2

P.O. Box 382612

Duncanville, TX 75138

USA

www.drkimberlyellison.com

To the absolute loves of my life and my true destiny pushers.

Faron L. Ellison, Jr., McCarley Nichole, McKinley Grace and Faron L. Ellison, III.

Table of Contents

Acknowledgements

Vision is a strategic guide dedicated to every person who is pregnant with purpose and ready to enter the birthing room. To every person whose dreams make him or her afraid, this book is for you. It was designed with you in mind. This guide is a part of my Habakkuk 2:2-3 where the Lord gives the answer to the Prophet Habakkuk, instructing him to write the vision and make it plain, that he may run who reads it. You are the “he” who reads it. My prayer is that you take every part of this strategic trilogy and run; run towards your fears right into your destiny; making every giant you face know the power of the amazing God you serve. Let this book serve as your five stones and a sling, for there is a David that lives in you. Go slay the giant!

Preface

Vision is about becoming the grandest version of the greatest vision you've ever seen of yourself. It's about facing the mental, emotional, physical and financial giants of your life with the spiritual giant that lives on the inside. Vision is about living a life of destiny, promise and purpose unapologetically. This amazing strategy focused workbook, planner and journal is nothing short of vision made manifest. Being a visionary is a gift and often times the gift can feel like a curse. For those who share in this gift of seeing the unseen, you understand that vision is not a noun; it is a verb in its finest form. Vision, the book, is intended to bring structure to individuals whose thoughts travel the farthest distance from their vision. It is written to help create milestones that will lead to destiny locations along life's journey. Vision is my tablet, where I write the vision plain so that those who read it may run.

Vision is unlike anything you've read. It is truly designed to bring your vision to action. Since I was a young girl, I dreamed of speaking in front of hundreds of thousands of people, divinely empowering them to dream while living. I saw the life I live today. The vision I saw was the seed to the purpose and destiny that has manifested in my life. The key word here is "seed." Your vision is your seed however; it will not grow if you do not tend to its needs. Over the years, I've used a technique of seeing, writing, planning and praying to bring my goals and visions to life and now I am sharing with you the practical, yet intentional and purposeful proven strategy for bringing your vision to life. Consider yourself pregnant and welcome to the birthing room!

Introduction

Vision is the seed to your purpose and the pathway to destiny however, without bringing your vision to life you are merely living a nightmare. Can you imagine what it feels like to live a life without vision or a life without a clue as to how to make your vision a part of your reality? Vision is a strategic trilogy designed to help you bring your vision to life. With all of life's demands and competing priorities it is almost impossible to accomplish anything that requires a strategy. This action packed guide is comprised of three critical resources to help you bring to reality the images you cut and paste onto your annual vision board. The strategic focus of Vision provides you a vision to action approach that teaches you how to write, plan and pray.

Part One: Write

Part one of the Vision to Action Strategy is WRITE. This is the part of the strategy where the work takes place. It is critical to think through your vision and begin to capture what you see. This is considered the workbook, the part of the plan where you write the vision and make it plain. Be prepared to put in the work.

Part Two: Plan

Part two of the Vision to Action Strategy is PLAN. This is the part of the strategy where your written vision moves from Vis10N to Action using the 10 Point Plan designed to take your vision through a life cycle of expectancy. This is considered the strategic action plan. Be prepared to put in the work.

Part Three: Prayer

Part three of the Vision to Action Strategy is Pray. This is the part of the strategy where you cover your written vision and 10 Point Plan in prayer. This is considered the prayer journal. Be prepared to go to war.

This strategic trilogy was designed in order for you to become the grandest version of the greatest vision you've ever seen of yourself. Vision awaits.

Part One

WRITE. Plan. Pray.

Part One: Write

Then the Lord answered me and said: "Write the vision and make it plain on tablets, that he may run who reads it. For the vision is yet for an appointed time; but at the end it will speak, and it will not lie. Though it tarries, wait for it; because it will surely come, it will not tarry."

- Habakkuk 2:2-3

Part one of the Vision to Action Strategy is WRITE. One of the most common beliefs yet uncommon practices is for individuals to write down their vision. Let's be clear. In Habakkuk 2:2-3, the Lord provided an answer to the Prophet Habakkuk. He provided an answer without assumption. In order for the vision to manifest, Habakkuk had to write it down. Not so much to remember, but more so to clearly define what the vision revealed to him. The vision had to be written not only for Habakkuk, but also for the impact and purpose the vision would have on others. So writing the vision is only the beginning.

Habakkuk was also instructed to make it plain; thus making it where anyone who reads the vision will have a pure understanding. This scripture alone lets you know your vision is not just for you. It further confirms that vision must be brought into action. The action begins with you writing your vision down. In part one of this strategic plan, your role is simple. You are to write down everything you see about your future. Your goal is to capture the very essence of who you are in the vision. If your vision about the vision is unclear or blurry, no worry, part one is also designed to help you remember what was forgotten over time.

One thing to keep in mind is that bringing your vision to action will take time and effort however, it's not about how fast you finish rather the purpose is to bring life to your vision, in order for the power and purpose locked up on the inside of you manifests. This is not a novel, it is a three part strategic plan designed to bring your vision to action.

Let the work begin. Let's WRITE!

What is Vision?

1) The faculty or state of being able to see.

   "She had defective vision."

   Synonyms: eyesight, sight, observation, (visual) perception

2) An experience of seeing someone or something in a dream or trance, or as a supernatural apparition.

   "The idea came to him in a vision."

   Synonyms: apparition, hallucination, illusion, mirage, phantom, ghost, wraith, and manifestation

Verb rare

1) Imagine.

Mentions of Vision in the Bible

In the bible vision is mentioned 79 times and in 73 different verses with 36 specific bible verses about vision.

Whether it is a vision someone saw or experienced it was followed up with a call to action or something happening or had happened.

- Warning
- Take heed
- Take action
- Spoke to a future state of being

There is a difference between having visions and having a vision…

1) **Having a vision** – Being able to see
2) **Having visions** – Having an experience

You must begin to experience what you see. Vision to action is having visions about your vision; it's vision twice. Seeing the vision and experiencing or becoming the vision.

Have you envisioned yourself doing something or being a part of something that has not happened yet? If so, what is it?

____

____

____

____

What Do You See?

1) What do you see?

2) What do you see yourself doing?

3) What version of the vision do you see today?

Vision Statement vs. Perception Statement

Creating your vision statement will chronicle the details of what you see, however, your perception statement will determine how you live out the details. What lies behind the eye is greater than what lies before it. In this context, seeing is not believing. Moving your vision to action requires you to change your self-perception and personal belief system.

1) What is your vision statement?

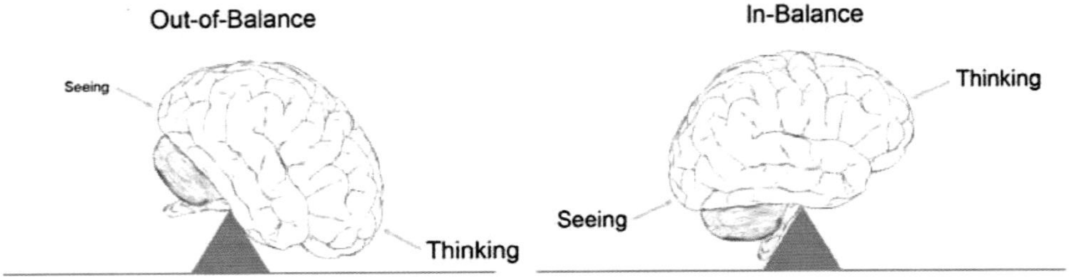

____

____

2) Now, what is your perception statement? What do you need to start, stop or continue thinking about your ability to experience your vision?

____

____

15 Questions to Discover Your Life's Purpose

The following are a list of questions that can assist you in discovering your purpose. They are meant as a guide to help you get into a frame of mind that will be conducive to defining your personal mission.

Simple Instructions:

- Find a place where you will not be interrupted. Turn off your cell phone.
- Write the answers to each question down. Write the first thing that pops into your head. Write without editing. It's important to write out your answers rather than just thinking about them.

- Write quickly. Give yourself less than 60 seconds per question. Preferably less than 30 seconds.
- Be honest. Nobody will read it. It's important to write without editing.
- Enjoy the moment and smile as you write.

15 Questions:

1) What makes you smile? (Activities, people, events, hobbies, projects, etc.)

____

____

2) What were your favorite things to do in the past? What about now?

____

____

3) What activities make you lose track of time?

____

____

4) What makes you feel great about yourself?

____

____

5) Who inspires you most? (Anyone you know or do not know. (Family, friends, authors, artists, leaders, etc.) Which qualities inspire you in each person?

____

____

6) What are you naturally good at? (Skills, abilities, gifts, etc.)

____

____

7) What do people typically ask you for help in?

____

____

8) If you had to teach something, what would you teach?

____

____

9) What would you regret not fully doing, being or having in your life?

____

____

10) You are now 90 years old, sitting on a rocking chair on your porch; you can feel the spring breeze gently brushing against your face. You are blissful and happy and are pleased with the wonderful life you've been blessed with. Looking back at your life and all that you've achieved and acquired, all the relationships you've developed; what matters to you most? List them out.

____

____

11) What are your deepest values?

____

____

12) What were some challenges, difficulties, and hardships you've overcome or are in the process of overcoming? How did you do it?

____

____

13) What causes do you strongly believe in? Connect with?

____

____

14) If you could get a message across to a large group of people, who would those people be? What would your message be?

____

____

15) Given your talents, passions and values, how could you use these resources to serve, to help, to contribute? (To people, beings, causes, organizations, environment, planet, etc.)

____

____

What Are Your Highest Values?

Please select 3-5, and put in order most important to you:

Achievement
Adventure
Beauty
Being the best
Challenge
Comfort
Courage
Creativity
Curiosity
Education
Empowerment
Environment
Family
Financial freedom
Fitness

Friendship
Giving I service
Health
Honesty
Independence
Inner peace
Integrity
Intelligence
Intimacy
Joy
Leadership
Learning
Love
Motivation
Passion

Performance
Personal growth
Play
Productivity
Primary
Relationship
Reliability
Respect
Security
Spirituality
Success
Time freedom
Variety

____

Others Values Not Listed:

Your Personal Mission Statement

"Writing or reviewing a mission statement changes you because it forces you to think through your priorities deeply, carefully, and to align your behavior with your beliefs."

~Stephen Covey

A personal mission consists of 3 parts:

- What do I want to do?
- Who do I want to help?
- What is the result? What value will I create?

Steps to Creating Your Personal Mission Statement:

1) Do the exercise with the 15 questions above as quickly as you can.

____

____

____

2) List out action words you connect with.

Example: educate, accomplish, empower, encourage, improve, help, give, guide, inspire, integrate, master, motivate, nurture, organize, produce, promote, travel, spread, share, satisfy, understand, teach, write, etc.

____

____

____

3) Based on your answers to the 15 questions, list everything and everyone that you believe you can help.

   Example: People, creatures, organizations, causes, groups, environment, etc.

   ____

   ____

   ____

4) Identify your end goal. How will the 'who' from your answer above benefit from what you 'do?'

   ____

   ____

   ____

5) Combine steps 2-4 into a sentence, or 2-3 sentences.

   ____

   ____

   ____

Example: To transform, inspire and empower women and young girls to live out their purpose, know their value, overcome societal, generational and mental barriers in order to grow and become successful in the personal, professional and spiritual areas of their life. – **Dr. Kimberly Ellison, Personal Mission Statement**

Adopted from: http://thinksimplenow.com/happiness/life-on-purpose-15-questions-to-discover-your-personal-mission

Casting the Vision: Goal Alignment

The term "casting" means to make a mold, form or manufacture. In order for your vision to come to life and take form, you must cast it. Creating your goals are the tablets spoken of in Habakkuk 2:2, *"And the LORD answered me: "Write the vision; make it plain on tablets, so he may run who reads it."* Your vision is what you see, your vision statement is what you write down, your goals make the vision plain for you and enable you and others to run (carry out) your vision.

Vision (Write down)

- Drives Belief
- Future Focused
- Purpose Driven

Goals (Make it plain)

- Determines Strategy
- Drives Process
- Develops Plan

Action (Run)

- Moves You Closer to the Vision
- Creates Action
- Measureable
- Tangible
- Others grow from it

Now that you know what it means to write the vision and make it plain, in the space below, identify 1-2 goals that support your vision statement. Go ahead and Write it!

Goal #1

____

____

____

____

____

____

Goal #2

____

____

____

____

____

____

Part Two

Write. **PLAN**. Pray.

Vis1On To Action

10 Point Plan

Part Two: PLAN

A simple man believes anything, but a prudent man gives thought to his steps."

- Proverbs 14:15

Part two of the Vision to Action Strategy is PLAN. By principle and by example, God's Word establishes strategic planning as one of the ways He works in and through His people. There are a number of leaders in the bible who thought and acted strategically, scripture indicates several individuals who accomplished seemingly difficult assignments with a well laid out plan. Even David had a plan to slay Goliath; it was well thought out and flawlessly executed. Nehemiah was a leader who used a strategy. When God laid it on his heart to rebuild the walls of Jerusalem, Nehemiah began to establish and then work through a well-planned strategy to accomplish the vision God had given. He assessed the damage. He secured the resources. He established leaders and distributed the assignments among them and there was great success.

Your vision or goal may not be as elaborate as building a wall like Nehemiah or slaying a giant like David did; however I do believe you will experience some walls and giants along your journey. Part two of the strategic trilogy provides a 10 Point Plan to help you bring your vision to life. By implementing this strategic and targeted approach you are sure to see your vision twice, once in your mind and again in real life! Based on where you are along your journey feel free to tweak the points as necessary however, do not abandon the plan.

Point 1: Write the Vision

Putting your vision on paper provides a sense of responsibility and accountability.

Point 2: Analyze Your Current Situation

In order to move from vision to action you must do a self-assessment; you must analyze and assess your level of readiness; your initial commitment and level of investment to your vision.

Point 3: Establish P2+S Goals

Identify your personal, professional and spiritual goals for your vision; identifying goals for each area gives you a targeted approach to accomplishing your goal in smaller steps.

Point 4: Determine Barriers to Entry

Barriers to entry are the existence of high startup costs or other obstacles that prevent new competitors from easily entering an industry or area of business. As it relates to your personal, professional and spiritual goals, the barriers to entry are called "distractions."

Point 5: Develop Strategies

Developing a strategic plan to accomplish your personal, professional or spiritual goals create a path toward implementation and completion.

Point 6: Resources Needed

A resource is a source or supply from which benefit is produced.

Point 7: Implement Strategies

Just like faith without works is dead so is vision without action. This step is where you execute the strategic plans.

Point 8: Evaluate Progress

Evaluating your progress is more than calculating profit or obtaining a position.

Point 9: Celebrate

Celebrating milestones are essential to the process of seeing your vision manifest.

Point 10: Dream Again

Dreaming again requires you to muster up the courage to risk it all, again!

VIS1ON TO ACTION
10 POINT PLAN
01
Write the Vision
02
Analyze Your Current Situation
03
Establish P2+S Goals
04
Determine Barriers to Entry
05
Develop Strategies
06
Resources Needed
07
Implement Strategies
08
Evaluate Progress
09
Celebrate
10
Dream Again
V2A
DR. KIMBERLY ELLISON
The Signature of Excellence K.E.
Bring Your Vision Board to Life with
Dr. Kimberly Ellison's V2A stategic plan!
www.drkimberlyellison.com

1) Where in this process do you need the most help?

____

____

____

2) What are ways to overcome that step in the process?

____

____

____

3) How will you sustain your approach?

____

____

____

Point 1: Write the Vision

"If you limit your choices only to what seems possible or reasonable, you disconnect yourself from what you truly want, and all that is left is a compromise."

- Robert Fritz

Putting your vision on paper provides a sense of responsibility and accountability. You are the only person who is required to show up for your vision board party! You are the only person required to describe what you see, feel, think and believe about your vision. Point 1 is where the magic happens it's where you put into words the picture you see in your mind; the feeling you get when you think about the possibility of your vision coming to life.

Below, begin to brainstorm your vision. Put your vision into words by describing what you see? What is your vision description? Paint a graphic mental picture of your vision.

____

____

____

____

____

Rewrite your vision statement below. How does your description and statement align with what you see in your future?

____

____

____

____

____

Point 2: Analyze Your Current Situation

"Create your future from your future, not your past."

- Werner Erhard

In order to move from vision to action you must do a self-assessment; you must analyze and assess your level of readiness; your initial commitment and level of investment to your vision. How much time do you plan to invest in bringing your vision to action? Assess your level of readiness by your current season. Point two, Analyzing Your Current Situation helps you to prepare for the anticipated distractions, commitment, financial investment, time investment, resource investment, family investment and self-investment.

Changing anything for any reason takes take time, energy and effort. You must assess your capacity to create and invest.

Describe the current season of your life.

What do you need to:	Describe the action.	By what date?
Stop		
Start		
Continue		

Let's Take a Selfie!

When taking a selfie it's important to know your good side and your bad angles. Often times we have to take numerous pictures to find the perfect selfie. Modern day technology has gifted us with filters and effects that will allow us to change the raw photo, which in turn impacts the true quality of the image. In order to move your vision to action you must see yourself from the inside, out. It's important to know your strengths and weaknesses; your areas of opportunities.

In terms of preparing to move your vision forward what are your strengths? What skills, gifts, talents, and experience do you have that will help you move your vision forward?

____

____

____

____

____

What are your areas of opportunities? What are your gaps in knowledge, skill, or behavior?

____

____

____

____

____

Point 3: Establish Goals P2+S

"The best way to predict the future is to create It."

- Alan Kay

Goal setting is truly underrated. Establishing your goals is equivalent to an Olympic track competitor getting on their mark. Your goals prepare you for the journey ahead. They keep you focused on the purpose, the finish line and the outcomes. It's important to identify your personal, professional and spiritual goals for your vision; identifying goals for each area gives you a targeted approach to accomplishing your goal in smaller steps.

The purpose is not to just put words on paper, the purpose is to capture S.M.A.R.T. goals or shall we say, "S.M.A.R.T.E.R." goals.

1) Specific & Simple
2) Measurable & Meaningful
3) Attainable & Achievable
4) Relevant & Results – Oriented
5) Timely & Trackable
6) Evaluate & Effective
7) Re-evaluate & Reward

Create S.M.A.R.T.E.R. personal, professional and spiritual goals.

What goal do you need to get S.M.A.R.T.E.R about? Write the goal down below and make it S.M.A.R.T.E.R.

Goal: ____

S: ____

M: ____

A: ____

R: ____

T: ____

E: ____

R: ____

Point 4: Determine Barriers to Entry

"It's not enough to be busy, so are the ants. The question is, what are we busy about?"

- Henry David Thoreau

Barriers to entry are the existence of high startup costs or other obstacles that prevent new competitors from easily entering an industry or area of business. As it relates to your personal, professional and spiritual goals, the barriers to entry are called **"distractions."** Consider your barriers to entry. What people, places or things do you feel are distractions? Some distractions are blatant while others are subtle. Blatant distractions are easy to identify and can oftentimes be better maintained while subtle distractions are more difficult to recognize and often go undetected.

Examples of Blatant distractions:

- *Television*
- *Family*
- *Kids*
- *Relationships*
- *Cell Phone*

Identify a few more obvious distractions that could serve as your barriers to entry.

1) ____

____

____

2) ____

____

____

3) ____

____

____

Examples of Subtle distractions:

- *Your thought life*
- *Lack of prioritizing your time*
- *Social Media*
- *Procrastination*
- *Over committing yourself*

Identify a few more subtle distractions that could serve as your barriers to entry.

1) ____

____

____

2) ____

____

____

3) ____

____

____

Point 4: Determine Barriers to Entry (*continued)*

"Whatever you can do or dream you can, begin it. Boldness has genius, and magic and power in it. Begin it now."

– Goethe

A barrier to entry is something that blocks or impedes the ability of an individual or company (competitor) to enter an industry. For those who have an entrepreneurial spirit or desire to establish a non-profit organization, economic barriers to entry often affect your vision, passion and creativity. Well, in order to move past the barriers you must first identify and understand the direct threat and or impact they bring to your vision.

Determine the current impact of the economic barrier you are facing. What are upfront costs to start?

____

____

____

____

____

What personal luxury or habit are you willing to sacrifice in order to work toward your upfront cost?

____

____

____

____

____

How have the barriers to entry impacted your forward mobility?

____

____

____

____

____

Point 5: Develop Strategies

Vision without action is a daydream. Action without vision is a nightmare."

- Japanese Proverb

A strategy is a way of describing how you are going to get things done. Developing a strategic plan to accomplish your personal, professional or spiritual goals create a path toward implementation and completion. Keep in mind your strategy is your vision written as action steps into manageable size pieces. As you write your strategic plan you begin to see your vision manifest on paper and in real life. Your strategies should directly correlate with your P2+S goals.

Place one of your goals from Point 3 below.

Ex. Start my own business; Ex. become more active in church

I will pursue the following strategies to reach the goal identified in the statement above.

Ex. File for LLC or C-Corp by Dec. 31st; Ex. Join church ministry group

____

____

____

____

____

Point 6: Resources Needed

"The future belongs to those who see possibilities before they become obvious."

\- John Scully

A resource is a source or supply from which benefit is produced. Typically resources are materials, energy, services, staff, knowledge, or other assets that are transformed to produce benefit and in the process may be consumed or made unavailable. In moving from vision to action you must determine the type of resources needed to produce the outcome you desire. Whether your goal is personal, professional or spiritual, you will need people, places or things to accomplish them.

Here are 5 types of resources to consider when moving from vision to action.

Financial Resources

- What type and how much money do you need? Financial resources can be obtained from a variety of sources based on the type of goal and/or vision.

Human Resources

- No one can go at it alone. Who do you need to help bring your vision to action?

Educational Resources

- How much do you know about what you're trying to accomplish? Do you have enough information and knowledge about particulars associated with your goal and/or vision?

Physical Resources

- Are there any physical resources needed to support your goal and/or vision? Do you need a building, location, computers, etc.

Emotional Resources

- What type of support system do you need? Identify a few people to encourage you when times become stressful, overwhelming and you want to give up. It's important to also have a support team who will cheer you on and celebrate every win with you!

Point 7: Implement Strategies

"People only see what they are prepared to see."

- Ralph Waldo Emerson

Just like faith without works is dead, so is vision without action. This step is where you execute the strategic plans you've identified in Point 5. By now, you have strategically laid out your plan of action and should be in position to fully carry out each action. During implementation you may have to course correct as you move forward. Often times, well laid plans need to be adjusted due to unknown, hidden or new circumstances. Here is where you lay all fear, doubt and uncertainty to rest and become the best version of the vision you've ever dreamed for yourself.

Point 7 Implementation Strategies is the answer to the question, "Where are you going?" This point is your compass and vision roadmap needed for every step along your journey. Expect just about everything you mentioned in Point 4, Barriers to Entry to start showing up. Yes, every distraction possible will come to distort your vision and derail your goals.

Here are a few ways to "distraction proof" your implementation strategies.

1) Reread your written vision (Point 1)
2) Remember your strengths and opportunities (Point 2)
3) Revisit your P2+S goals (Point 3)
4) Recall your blatant and subtle distractions (Point 4)

5) Remind yourself of the strategic plan (Point 5)

6) Rely on your emotional resources (Point 6)

Remember, this Vis1ON to Action is a strategic plan to help bring your vision to life. Use each point in the plan to move your vision to action. Implementation is critical to the purpose, promise and destiny on your life.

Point 8: Evaluate Progress

"If one advances confidently in the direction of his dreams, and endeavors to live the life which he has imagined, he will meet with success unexpected in common hours."

- Henry David Thoreau

Point 8 Evaluate Progress is a vision must have. In order to move forward in living a life of destiny, promise and purpose to bring your vision to action; evaluation of your progress is an essential point of the strategic plan. It is necessary and needed in order to continue your journey. Evaluating your progress is more than calculating profit or obtaining a position. It is truly relative to living the life you were destined to live. Are you living the vision God showed you? Are you becoming the person of faith, the business owner, the mother, the wife, the leader, or woman you were created to become?

As it relates to the 10 Point Plan you go back through each point and evaluate what worked well and what did not. However, your evaluation should bring a sense of accomplishment and achievement as you took the risk to live on purpose! You stepped outside of your comfort zone and put your plan into action.

Here are a few ways to evaluate your progress:

1) Ensure alignment of P2+S goals with outcomes.

2) Review results of outcomes

3) Is there an immediate Return on Investment (ROI) (i.e. relationships, financial growth, increased clientele, stronger personal relationship with God, etc.)?
4) Are my actions moving me in the right direction?
5) Do I still desire to accomplish my goal(s) and/or pursue my vision?
6) What do I need to change about my vision to action plan?
7) What other resources do I need?
8) How is my level of commitment in moving forward?
9) Have my original objectives/goals changed?
10) What distractions do I need to remove?

In order to successfully evaluate your progress it's imperative that you do an evaluation or quality check throughout your journey. Remember, you can course correct as you see necessary to accomplish your goal and bring your vision to action.

Point 9: Celebrate

"Vision animates, inspires, and transforms purpose into action."

- Warren Bennis

Celebrating milestones are essential to the process of seeing your vision manifest. Often times we focus so much on the process that we forget to celebrate the journey. Celebrating big or small successes throughout the process can often help build your confidence and allow you to see the vision unfold. One simple way to celebrate is to reward yourself for accomplishing action items on your daily to-do list. Remember, slow progress is progress. The closer you are to accomplishing your goal the more in focus your vision becomes. Begin to celebrate weekly, monthly, quarterly and annual milestones. Your vision is worth the sacrifice! Your goals are worth the discipline and you surely deserve to celebrate your commitment. Learn to celebrate YOU!

Here are a few ways to celebrate bringing your vision to action.

1) Take a short break away from working toward the goal/vision. Celebrate by doing something that you typically would feel guilty about spending your time.

2) Grab your favorite dessert and enjoy it alone. Don't share with anyone, not even the kids.

3) Treat yourself to your favorite restaurant and if your budget allows go watch a movie at the theatre. There are still dollar movie theatres in some cities.

4) Take a selfie with your completed to-do list.

5) Get dressed up with nowhere to go and celebrate you.

6) Send yourself flowers or chocolate covered strawberries.

7) Put on a new lipstick color.

8) Write in your journal how you feel in this very moment.

The biggest roadblocks in success are fear and self-doubt. Learning to celebrate you will help you replace fear with faith and self-doubt with "Godfidence," finding confidence in God. It's time to stop celebrating everyone else and include you in the party. It's time to bring your vision to action.

Point 10: Dream Again

"Cherish your visions and your dreams as they are the children of your soul, the blueprints of your ultimate achievements."

- Napoleon Hill

Often times we think once we have accomplished the goal or the vision has been made manifest, we can continue operating in the same season. Unfortunately, in order to have a bigger vision you must dream again. You must ask God for divine creativity and empowerment. Dreaming again requires you to muster up the courage to risk it all, again! It requires you to face the giant like David did. It requires you to pull out your little sling and small pebbles of hope to slay the Goliath in front of you. Dreaming again requires you to reset, refuel and re-ignite the passion and purpose for the next vision.

The Vis10n to Action 10 Point Plan is designed to help you bring your vision to life, it is designed to help you to move past the cute pictures in the magazine and become the grandest version of the greatest vision God has shown you! No more vision board parties without the action plan to bring your vision to life. You can no longer allow the imagery of the magazine images to fancy your imagination; it's time to live the true reality of what you see.

Part Three

Write. Plan. **PRAY**.

Part Three: PRAY

Do not be anxious about anything, but in every situation, by prayer and petition, with thanksgiving, present your requests to God.

- Philippians 4:6

Part three of the Vision to Action Strategy is PRAYER. There are several scriptures that could serve as anchors to your vision plan however, one that comes to mind is Proverbs 16:3, "Commit your actions unto the Lord and your plans will succeed." Whether you want to accomplish a small personal goal or an extremely large business goal, God is interested in partnering with you to bring it to life. As you write and plan do not forget to pray. Pray about next steps, financial decisions, resources and your process to purpose. As you walk out the steps of the Vis10n to Action *10 Point Plan* use this part of the strategy to learn more about the steps you are taking to bring your vision to reality.

The Vision Prayer Journal is a simple, yet personal tool to use for learning more about your vision and what God is saying and where He is leading you on your path to destiny. It is important to cover each point in prayer in order to partner with God through every point of the plan. Invest time in reading, praying, listening and acting on what God will download along your journey. The final part of this strategic trilogy is the most important, this is where you will receive everything you need to write and plan your vision.

See how the Vision Prayer Journal works.

1) There are four journal pages for ***each point in the*** Vis10n to Action 10 Point Plan.

    - Bible Study
    - Prayer Journal
    - Prayer Checklist
    - Hearing from God

2) As you journey through each point in the plan spend some time studying about that particular point in the plan and fill out the journal pages as God speaks to you through His word and His voice.

Point 1: Write the Vision - Bible Study

BIBLE VERSE

APPLICATION TO MY LIFE	DEFINITIONS

WHAT DID I LEARN HERE ABOUT MYSELF?	WHAT DID I LEARN HERE ABOUT GOD?

WHERE CAN I GROW? WHAT CHANGES CAN GOD MAKE IN MY LIFE?

Point 1: Write the Vision - Prayer Journal

Bible Reading:

How can I apply this passage to my life and how does this scripture speak to me?

I'M GRATEFUL FOR

MEMORY VERSE

MY PRAYER NEEDS

DEVOTIONAL TIME

ANSWERED PRAYERS

Point 1: Write the Vision - Prayer Checklist

DATE	PRAYER REQUEST	PROGRESS. ANSWERS. SCRIPTURES.	DONE

Point 1: Write the Vision - Hearing from God

Point 2: Analyze Your Current Situation - Bible Study

BIBLE VERSE

APPLICATION TO MY LIFE	DEFINITIONS

WHAT DID I LEARN HERE ABOUT MYSELF?	WHAT DID I LEARN HERE ABOUT GOD?

WHERE CAN I GROW? WHAT CHANGES CAN GOD MAKE IN MY LIFE?

Point 2: Analyze Your Current Situation - Prayer Journal

Bible Reading:

How can I apply this passage to my life and how does this scripture speak to me?

I'M GRATEFUL FOR

MEMORY VERSE

MY PRAYER NEEDS

DEVOTIONAL TIME

ANSWERED PRAYERS

Point 2: Analyze Your Current Situation - Prayer Checklist

DATE	PRAYER REQUEST	PROGRESS. ANSWERS. SCRIPTURES.	DONE

Point 2: Analyze Your Current Situation - Hearing from God

Point 3: Establish P2+S Goals - Bible Study

BIBLE VERSE

APPLICATION TO MY LIFE	DEFINITIONS

WHAT DID I LEARN HERE ABOUT MYSELF?	WHAT DID I LEARN HERE ABOUT GOD?

WHERE CAN I GROW? WHAT CHANGES CAN GOD MAKE IN MY LIFE?

Point 3: Establish P2+S Goals - Prayer Journal

Bible Reading:

I'M GRATEFUL FOR

MEMORY VERSE

MY PRAYER NEEDS

ANSWERED PRAYERS

How can I apply this passage to my life and how does this scripture speak to me?

DEVOTIONAL TIME

Point 3: Establish P2+S Goals - Prayer Checklist

DATE	PRAYER REQUEST	PROGRESS. ANSWERS. SCRIPTURES.	DONE

Point 3: Establish P2+S Goals - Hearing from God

Point 4: Determine Barriers to Entry - Bible Study

BIBLE VERSE

APPLICATION TO MY LIFE	DEFINITIONS

WHAT DID I LEARN HERE ABOUT MYSELF?	WHAT DID I LEARN HERE ABOUT GOD?

WHERE CAN I GROW? WHAT CHANGES CAN GOD MAKE IN MY LIFE?

Point 4: Determine Barriers to Entry - Prayer Journal

Bible Reading:

I'M GRATEFUL FOR

MEMORY VERSE

MY PRAYER NEEDS

ANSWERED PRAYERS

How can I apply this passage to my life and how does this scripture speak to me?

DEVOTIONAL TIME

Point 4: Determine Barriers to Entry - Prayer Checklist

DATE	PRAYER REQUEST	PROGRESS. ANSWERS. SCRIPTURES.	DONE

Point 4: Determine Barriers to Entry - Hearing from God

Point 5: Develop Strategies - Bible Study

BIBLE VERSE

APPLICATION TO MY LIFE	DEFINITIONS

WHAT DID I LEARN HERE ABOUT MYSELF?	WHAT DID I LEARN HERE ABOUT GOD?

WHERE CAN I GROW? WHAT CHANGES CAN GOD MAKE IN MY LIFE?

Point 5: Develop Strategies - Prayer Journal

Bible Reading:

How can I apply this passage to my life and how does this scripture speak to me?

I'M GRATEFUL FOR

MEMORY VERSE

MY PRAYER NEEDS

DEVOTIONAL TIME

ANSWERED PRAYERS

Point 5: Develop Strategies - Prayer Checklist

DATE	PRAYER REQUEST	PROGRESS. ANSWERS. SCRIPTURES.	DONE

Point 5: Develop Strategies - Hearing from God

Point 6: Resources Needed - Bible Study

BIBLE VERSE

APPLICATION TO MY LIFE	DEFINITIONS

WHAT DID I LEARN HERE ABOUT MYSELF?	WHAT DID I LEARN HERE ABOUT GOD?

WHERE CAN I GROW? WHAT CHANGES CAN GOD MAKE IN MY LIFE?

Point 6: Resources Needed - Prayer Journal

Bible Reading:

How can I apply this passage to my life and how does this scripture speak to me?

I'M GRATEFUL FOR

MEMORY VERSE

MY PRAYER NEEDS

DEVOTIONAL TIME

ANSWERED PRAYERS

Point 6: Resources Needed - Prayer Checklist

DATE	PRAYER REQUEST	PROGRESS. ANSWERS. SCRIPTURES.	DONE

Point 6: Resources Needed - Hearing from God

Point 7: Implement Strategies - Bible Study

BIBLE VERSE

APPLICATION TO MY LIFE	DEFINITIONS

WHAT DID I LEARN HERE ABOUT MYSELF?	WHAT DID I LEARN HERE ABOUT GOD?

WHERE CAN I GROW? WHAT CHANGES CAN GOD MAKE IN MY LIFE?

Point 7: Implement Strategies - Prayer Journal

Bible Reading:

How can I apply this passage to my life and how does this scripture speak to me?

I'M GRATEFUL FOR

MEMORY VERSE

MY PRAYER NEEDS

ANSWERED PRAYERS

DEVOTIONAL TIME

Point 7: Implement Strategies - Prayer Checklist

DATE	PRAYER REQUEST	PROGRESS. ANSWERS. SCRIPTURES.	DONE

Point 7: Implement Strategies - Hearing from God

Point 8: Evaluate Progress - Bible Study

BIBLE VERSE

APPLICATION TO MY LIFE	DEFINITIONS

WHAT DID I LEARN HERE ABOUT MYSELF?	WHAT DID I LEARN HERE ABOUT GOD?

WHERE CAN I GROW? WHAT CHANGES CAN GOD MAKE IN MY LIFE?

Point 8: Evaluate Progress - Prayer Journal

Bible Reading:

I'M GRATEFUL FOR

MEMORY VERSE

MY PRAYER NEEDS

ANSWERED PRAYERS

How can I apply this passage to my life and how does this scripture speak to me?

DEVOTIONAL TIME

Point 8: Evaluate Progress - Prayer Checklist

DATE	PRAYER REQUEST	PROGRESS. ANSWERS. SCRIPTURES.	DONE

Point 8: Evaluate Progress - Hearing from God

Point 9: Celebrate - Bible Study

BIBLE VERSE

APPLICATION TO MY LIFE	DEFINITIONS

WHAT DID I LEARN HERE ABOUT MYSELF?	WHAT DID I LEARN HERE ABOUT GOD?

WHERE CAN I GROW? WHAT CHANGES CAN GOD MAKE IN MY LIFE?

Point 9: Celebrate - Prayer Journal

Bible Reading:

How can I apply this passage to my life and how does this scripture speak to me?

I'M GRATEFUL FOR

MEMORY VERSE

MY PRAYER NEEDS

ANSWERED PRAYERS

DEVOTIONAL TIME

Point 9: Celebrate - Prayer Checklist

DATE	PRAYER REQUEST	PROGRESS. ANSWERS. SCRIPTURES.	DONE

Point 9: Celebrate - Hearing from God

Point 10: Dream Again - Bible Study

BIBLE VERSE

APPLICATION TO MY LIFE	DEFINITIONS

WHAT DID I LEARN HERE ABOUT MYSELF?	WHAT DID I LEARN HERE ABOUT GOD?

WHERE CAN I GROW? WHAT CHANGES CAN GOD MAKE IN MY LIFE?

Point 10: Dream Again - Prayer Journal

Bible Reading:

How can I apply this passage to my life and how does this scripture speak to me?

I'M GRATEFUL FOR

MEMORY VERSE

MY PRAYER NEEDS

DEVOTIONAL TIME

ANSWERED PRAYERS

Point 10: Dream Again - Prayer Checklist

DATE	PRAYER REQUEST	PROGRESS. ANSWERS. SCRIPTURES.	DONE

Point 10: Dream Again - Hearing from God

The Vision to Action Journal

PRACTICE.

Moving from Vision to ACTION

Are you ready to work toward accomplishing your personal, professional and spiritual goals? If so, this Vision to Action Journal is for you. What is it going to take for you to move from vision to action? Through this customized action-planning journal you will watch your goals become results and your vision a reality. Please use the remaining pages of this guide to create a strategic plan of action for your goals throughout the year. The best way to accomplish the process is to practice the process. All of God's very best to you along your journey.

Move Your Vision 2 Action

Example: Goals, Goals, and more Goals…

Personal Goal:
To de-clutter every room in my home within the next 3 months.

Supporting Scripture:
Moreover, it is required of stewards that they are found trustworthy. 1 Corinthians 4:2

Affirmation Statement:
I am capable of removing unnecessary clutter from my life. Anything that is not being used is taking up space, not adding value and causing me to live a thought cluttered lifestyle. I am able to declutter my home at my pace over the next 3 months.

Strategies: Activities to Achieve Goal	Action Steps: What have you done to put this action in motion?	Target Date	Actual Date	Reflection
Purchase Containers	Reviewed Pinterest; Asked friends who are great organizers	12/19		This process was a bit overwhelming and I over promised on doing this action.
Decide on what room to do first	Taken inventory of each room; Taken photos of current clutter	12/26		I enjoyed this process. I really wasn't aware of all the things I've held on to.
Identify a process to give away items or throw away.	Purchased 39 -gallon trash bags. Marked calendar for bulk Monday. Solicited help from Faron regarding large items and trash.	01/19		I realized I have to be more specific in my request to remove bags and items. I will find a local charity to come pick up items.
Take photos of de-cluttered rooms	Taken photos	02/01		This was easy… I even posted to social media.
Put de-clutter safeguards in place.	Implemented a no longer needed or use policy. Put trash bins upstairs and downstairs.	03/19		I am not sure how to do this one just yet…but I have asked for help from friends that are organizers.

List of resources needed to accomplish goal:

Help to move large items and finances to purchase con

Time to Practice, Practice, Practice…

Personal/Professional/Spiritual Goal:
Supporting Scripture:
Affirmation Statement:

Strategies: Activities to Achieve Goal	Action Steps: What have you done to put this action in motion?	Target Date	Actual Date	Reflection

List of resources needed to accomplish goal:

Jewel Notes

Time to Practice, Practice, Practice...

Personal/Professional/Spiritual Goal:				
Supporting Scripture:				
Affirmation Statement:				
Strategies: Activities to Achieve Goal	**Action Steps: What have you done to put this action in motion?**	**Target Date**	**Actual Date**	**Reflection**

List of resources needed to accomplish goal:

Jewel Notes

Time to Practice, Practice, Practice…

Personal/Professional/Spiritual Goal:				
Supporting Scripture:				
Affirmation Statement:				
Strategies: Activities to Achieve Goal	**Action Steps: What have you done to put this action in motion?**	**Target Date**	**Actual Date**	**Reflection**

List of resources needed to accomplish goal:

Jewel Notes

Time to Practice, Practice, Practice…

Personal/Professional/Spiritual Goal:				
Supporting Scripture:				
Affirmation Statement:				
Strategies: Activities to Achieve Goal	**Action Steps: What have you done to put this action in motion?**	**Target Date**	**Actual Date**	**Reflection**

List of resources needed to accomplish goal:

Jewel Notes

Time to Practice, Practice, Practice…

Personal/Professional/Spiritual Goal:
Supporting Scripture:
Affirmation Statement:

Strategies: Activities to Achieve Goal	Action Steps: What have you done to put this action in motion?	Target Date	Actual Date	Reflection

List of resources needed to accomplish goal:

Jewel Notes

Time to Practice, Practice, Practice…

Personal/Professional/Spiritual Goal:				
Supporting Scripture:				
Affirmation Statement:				
Strategies: Activities to Achieve Goal	**Action Steps: What have you done to put this action in motion?**	**Target Date**	**Actual Date**	**Reflection**

List of resources needed to accomplish goal:

Jewel Notes

Time to Practice, Practice, Practice…

Personal/Professional/Spiritual Goal:				
Supporting Scripture:				
Affirmation Statement:				
Strategies: Activities to Achieve Goal	**Action Steps: What have you done to put this action in motion?**	**Target Date**	**Actual Date**	**Reflection**

List of resources needed to accomplish goal:

Jewel Notes

Time to Practice, Practice, Practice…

Personal/Professional/Spiritual Goal:				
Supporting Scripture:				
Affirmation Statement:				
Strategies: Activities to Achieve Goal	**Action Steps: What have you done to put this action in motion?**	**Target Date**	**Actual Date**	**Reflection**

List of resources needed to accomplish goal:

Jewel Notes

Time to Practice, Practice, Practice…

Personal/Professional/Spiritual Goal:				
Supporting Scripture:				
Affirmation Statement:				
Strategies: Activities to Achieve Goal	**Action Steps: What have you done to put this action in motion?**	**Target Date**	**Actual Date**	**Reflection**

List of resources needed to accomplish goal:

Jewel Notes

Time to Practice, Practice, Practice…

Personal/Professional/Spiritual Goal:
Supporting Scripture:
Affirmation Statement:

Strategies: Activities to Achieve Goal	Action Steps: What have you done to put this action in motion?	Target Date	Actual Date	Reflection

List of resources needed to accomplish goal:

Jewel Notes

Time to Practice, Practice, Practice…

Personal/Professional/Spiritual Goal:				
Supporting Scripture:				
Affirmation Statement:				
Strategies: Activities to Achieve Goal	**Action Steps: What have you done to put this action in motion?**	**Target Date**	**Actual Date**	**Reflection**

List of resources needed to accomplish goal:

Jewel Notes

Time to Practice, Practice, Practice…

Personal/Professional/Spiritual Goal:				
Supporting Scripture:				
Affirmation Statement:				
Strategies: Activities to Achieve Goal	**Action Steps: What have you done to put this action in motion?**	**Target Date**	**Actual Date**	**Reflection**

List of resources needed to accomplish goal:

Jewel Notes

Time to Practice, Practice, Practice…

Personal/Professional/Spiritual Goal:				
Supporting Scripture:				
Affirmation Statement:				
Strategies: Activities to Achieve Goal	**Action Steps: What have you done to put this action in motion?**	**Target Date**	**Actual Date**	**Reflection**

List of resources needed to accomplish goal:

Jewel Notes

Time to Practice, Practice, Practice…

Personal/Professional/Spiritual Goal:
Supporting Scripture:
Affirmation Statement:

Strategies: Activities to Achieve Goal	Action Steps: What have you done to put this action in motion?	Target Date	Actual Date	Reflection

List of resources needed to accomplish goal:

Jewel Notes

Time to Practice, Practice, Practice...

Personal/Professional/Spiritual Goal:
Supporting Scripture:
Affirmation Statement:

Strategies: Activities to Achieve Goal	Action Steps: What have you done to put this action in motion?	Target Date	Actual Date	Reflection

List of resources needed to accomplish goal:

Jewel Notes

Time to Practice, Practice, Practice…

Personal/Professional/Spiritual Goal:
Supporting Scripture:
Affirmation Statement:

Strategies: Activities to Achieve Goal	Action Steps: What have you done to put this action in motion?	Target Date	Actual Date	Reflection

List of resources needed to accomplish goal:

Jewel Notes

Time to Practice, Practice, Practice…

Personal/Professional/Spiritual Goal:
Supporting Scripture:
Affirmation Statement:

Strategies: Activities to Achieve Goal	Action Steps: What have you done to put this action in motion?	Target Date	Actual Date	Reflection

List of resources needed to accomplish goal:

Jewel Notes

Time to Practice, Practice, Practice…

Personal/Professional/Spiritual Goal:				
Supporting Scripture:				
Affirmation Statement:				
Strategies: Activities to Achieve Goal	**Action Steps: What have you done to put this action in motion?**	**Target Date**	**Actual Date**	**Reflection**

List of resources needed to accomplish goal:

Jewel Notes

Vision Made Manifest

Congratulations on the birth of your vision! Your hard work, dedication, commitment, sacrifice, tears and prayers have paid off. There is nothing like going through a strategic and tedious process and seeing the manifestation of your work for His glory! Think about all of the people who are now able to see, understand and run with your vision. It is because of your commitment to the process that you are able to experience destiny fulfilled. Now refer back to Point 10 of the Vis10n to Action Plan, it's time to start dreaming again.

vision

Connect with Dr. Kimberly Ellison On Social Media

Twitter: @drkellison

Instagram: @drkimberlyellison

Facebook: Dr. Kimberly Ellison

Contact Dr. Kimberly to schedule your **Vision to Action Intensive** Today! Email us at info@drkimberlyellison.com

FOR BOOKING, CONTACT US:

www.houseofruthjewels.com

www.drkimberlyellison.com

References

- Covey, S. R. (2004). *The 7 habits of highly effective people: Restoring the character ethic* ([Rev. ed.].). New York: Free Press.
- "Resource". En.wikipedia.org. N.p., 2016. Web. 22 Nov. 2016.
- Su, Tina. "Life On Purpose: 15 Questions To Discover Your Personal Mission". Thinksimplenow.com. N.p., 2016. Web. 22 Sept. 2015.
- *The Heart of Innovation: 50 Awesome Quotes on Vision. (2016).*
- Ideachampions.com. Retrieved 24 September 2016, from http://www.ideachampions.com/weblogs/archives/2011/01/50_awesome_quot_1.shtml

Made in United States
North Haven, CT
05 November 2026

21766168R60091